Wild Dogs and their Relatives

Contents

Cover: ***Golden jackal*** (Canis aureus) *Rhanthambhore, India* *Eric Dragesco/WWF*

Introduction 1

Fox cubs (Vulpes vulpes) *playing on an old tree stump.*

"Of the many ways of measuring a land's wealth, one of the surest signs of ecological richness and diversity is an abundance of predator species. Because each species sits at the top of a different food chain, belonging to a different cycle of organic matter, we can be certain of the existence of a larger animal community for every predator. This is, in turn, sustained by vegetation. The existence of carnivores carries the implications of a larger ecological community and of millions of years of evolutionary struggle."

The distinguished British wildlife biologist, Jonathan Kingdon, thus emphasises that conservation of the natural world, which is the foundation of all life on earth, must involve the predators. That they are not, themselves, always welcome neighbours and may have to be controlled in the public interest does not detract from their overall importance in the natural web of life. Predators contribute to the evolutionary health of their prey species by eliminating the weak and the sick. They are especially vulnerable to extinction because they range over large areas and live at low densities. As a result, efforts to remove predators from agricultural areas, or excessive exploitation of a species for commerce, may promote extinction of a species.

Grey wolf (Canis lupus) *with white-tailed deer kill, Montana, USA. Predators contribute to the evolutionary health of their prey species by eliminating the weak and the sick.*

Human attitudes to wild animals are varied. The fox may be viewed, simultaneously, as a resource by the fur trapper, a health hazard by the rabies official, a quarry by the huntsman, a subject by the photographer, vermin by the poultry farmer, and a work of art by the aesthete. Their judgments are neither right nor wrong, and have to be respected in planning conservation.

Although most wild dog species exist in reasonably healthy numbers, the African wild dog, red wolf and Simien jackal are threatened with extinction. Furthermore, bush dog, dhole, grey wolf, grey zorro, island grey fox and maned wolf could become endangered unless given more protection. More information about the status of nine other species is required in order to assess any threats to them. The remaining 16 of the

The African wild dog (Lycaon pictus) *is the most endangered large carnivore in Africa.*

34 species of wild dogs are not thought to need any immediate protective measures.

Recent surveys have indicated that the African wild dog is extinct in 19 out of 32 countries from which data have been collected. Only 6 countries - Botswana, Ethiopia, Kenya, Tanzania, Zambia and Zimbabwe - support wild dog populations which are likely to remain viable into the 21st century. But in 4 of these countries wild dogs have declined in numbers by at least one-third in the past decade, and little is known about the populations in Botswana and Ethiopia.

Their reputation as voracious killers of game and livestock has led to African wild dogs being heavily persecuted, and even to calls for their complete extermination. They are hunted and poisoned. Protection has been given only when they have reached low numbers. Diseases, such as rabies and anthrax, are further threats to the species. Because of their nomadic nature and large ranges, it is unlikely that African wild dogs can be conserved only in parks or reserves. Apart from rigorous protection, their conservation requires local people and farmers being educated to understand the ecological role of the wild dogs and discouraged from harassing and shooting them.

There are fewer than 1,000 surviving Simien jackals, all in the Ethiopian highlands. Most of them are in the Bale Mountains National Park and a single catastrophic event, such as an epidemic, precipitous reductions in prey species, or the changing climate, could wipe them out. These jackals also interbreed with domestic dogs, a threat to the survival of the species which could become particularly serious as populations become fragmented and reduced.

Improved conservation of the jackals in the Bale Mountains National Park could ensure their future survival. The remnant populations elsewhere may not survive. Established hybrids from breeding with domestic dogs need to be eliminated. The status of the jackals and their habitats needs continual monitoring to prevent deterioration. At the same time, a captive breeding programme is required as an assurance against extinction.

The red wolf in the United States became extinct in the wild during the 1970s as a result of a combination of factors including persecution, habitat loss, and hunting for its fur. However, a captive breeding programme provided 8 animals for reintroduction in 1988 in the Alligator River National Wildlife Refuge in North Carolina, where they are being carefully protected and monitored. It will take many years for a viable wild population to be established. ■

Red wolf (Canis rufus) *reintroduction programme, North Carolina, USA.*

Wild Dogs

The domestic dog is universally regarded as man's best friend. The many breeds include hunters and lap dogs, house pets and guard dogs, all loved for their loyalty to their masters and mistresses. Despite their extraordinary variation in form - more than any other domesticated animal - they are all descended from the wolf, which, paradoxically, has always been one of the most feared wild animals. No other species of wild dog appears to be an ancestor of the domestic breeds, even though many have been kept as pets.

This Welsh border collie 'Meg' became foster mother to a week-old fox cub. All modern varieties of domestic dog, universally regarded as man's best friend, have descended from the wolf.

From a scientific point of view, the evolution of the canid family is of major interest. The study of its adaptation to changing and challenging situations contributes to knowledge of evolution in general.

Scientists are not in agreement on the number of canid species. This booklet follows the classification of 34 species recognised by the Canid Specialist Group of The World Conservation Union.

The 34 species of wild dog belong to the family Canidae. It forms part of the Order of Carnivores or 'meat-eaters', together with the cats (Felidae), bears (Ursidae), raccoons (Procyonidae), weasels (Mustelidae), civets (Viverridae) and hyaenas (Hyaenidae). While these families look very different, they are grouped together because they all have molars, known as carnassial teeth, adapted for slicing flesh.

An Australian dingo (Canis familiaris dingo) *suckles her pups. The gestation period is about nine weeks in most dogs and there is only one litter a year.*

Carnivores are thought to be descended from small civet-like animals called 'miacids', which lived in the time of the dinosaurs. When the dinosaurs became extinct, the miacids evolved into the various families of carnivores. Each of these represents an adaptation to a characteristic way of life. Wild dogs, for example, pursue their prey, but cats stalk and ambush them. The cheetah, which is renowned for its swift running down of prey, is unique among cats in having several dog-like characteristics.

Whereas cats have flexible wrists, which enable them to grasp, dogs have fused wrist bones and locked radial and ulna bones in the forelimbs - adaptations helpful in running. Cats' claws are retracted, except when seizing prey, but those of dogs (and the cheetah) are always extended and leave marks in the footprint. All the dog species have long legs and lithe bodies, except for the South American bush dog and the raccoon dog, which are squat. Wild dogs have 5 toes (4 in the African wild dog), 42 teeth, except the bushdog (38), the bat-eared fox (48) and the Asiatic red dog or 'dhole' (40). Dhole females have 12-14 teats compared with 10 in other species.

Canids first evolved about 38 to 54 million years ago in the Eocene, and some ancient canids were long-bodied and short-limbed like civets. In the course of evolution, there were once 42 canid genera, but today there are just 10. The most primitive wild dog is the raccoon dog, an East Asian species introduced into the western USSR from 1927 onwards for fur-farming and now spreading through northern Europe. Its habit of hibernating is unique in the dog family.

The gestation period is about nine weeks in most dogs, and there is only one litter a year. Pups' eyes open at about two weeks. Solid food is taken after two to six weeks. Wolves, jackals, African wild dogs, and dholes regurgitate food for pups, while most other species carry it to the young.

Wild dogs vary in size almost as much as domestic dogs. The wolf, which may weigh up to 80kg, is the largest, while the fennec fox of the Arabian deserts is the smallest, usually weighing less than 1.5kg. Wild dogs have adapted to many types of habitat - Arctic wastes, cold northern forests, tropical dry and moist forests, grasslands and deserts. The larger species live mainly in packs, with a social hierarchy dominated by a leading male and a female. The wolf is a prime example of this characteristic, which is reflected in the way domestic dogs accept human families as their 'packs', in which they have their place and to which they are loyal. But most other wild dogs live in pairs, although some species, like jackals, may join up in larger groups.

Water does not deter wild dogs. They all swim well. The zorros of South America and the corsac fox of central Asia climb trees.

Wild dogs are found on all the continents except Australia and Antarctica. Australia has the dingo. It is now generally agreed that it was a domesticated species introduced by human settlers several thousand years ago. It became wild, or 'feral'. Nevertheless, biologists consider it an important species or subspecies. The red fox was taken to Australia by European settlers in the late 19th century.

Despite their affection for the domestic dog, people often view wild dogs as unwelcome neighbours. They are widely treated as vermin on the grounds that they attack livestock, from cattle and sheep to chickens, and may carry the deadly disease, rabies. Wolves and foxes are among 20 species that have been exploited for their fur. Historically several species have been severely threatened due to persecution, over-exploitation, and loss of habitat. ■

An African wild dog (Lycaon pictus) *stretches itself after cooling off in the river in South Luangwa National Park, Zambia.*

Wolves

The grey wolf once lived throughout most of Eurasia and North America when forests were widespread. It was the world's most widely distributed mammal. Today, it remains in healthy numbers only in Canada, Alaska, and the USSR, where the wolf population extends into Poland. Elsewhere, it has been exterminated, or survives in such small numbers that its future is in grave doubt.

The grey wolf (Canis lupus) *was once widely distributed through most of the Northern Hemisphere. It is now massively reduced in Europe and the United States of America.*

Largest of the wild dogs, the **grey** or **timber wolf** can grow a pelt varying in colour from nearly pure white and red or brown through to black. The largest specimens may weigh 80kg and be 150cm in length, with a tail of 31-51cm. They may reach 81cm at shoulder height. Females are smaller.

Wolves have a highly-developed social organisation in their packs, which are primarily family groups. Only the dominant male and female in a pack breed. The reproductive system of other females is suppressed, and any attempts by other pairs in the pack to mate will be broken up by the dominant pair. Packs may include up to 30 wolves, although they are usually smaller.

Wolves feed mainly on large ungulates, such as moose (known as elk in Europe), caribou (reindeer in Europe) and deer, but they will also eat smaller animals, as well as carrion and garbage. A wolf pack will test a herd by threatening it and picking out the weaker animals, which may be young, old or sick. Like other pack animals, they pursue their prey, snapping and biting it when close enough, and bringing it down. The kill is rapidly consumed.

Mating time is in winter, from January to April, and pups are born after 61-63 days. There may be up to 11 pups in a litter. After 8-10 weeks, pups are weaned. The whole pack shows great affection for the young, and pups may be adopted by another breeding female if the mother disappears. Sexual maturity usually comes at about 22 months, sometimes earlier, and a wild wolf may live for as long as 13 years.

The wolf's propensity to kill livestock, which it considers easy natural prey, has been its downfall. A measure of the toll is that an estimated one million cattle were killed by wolves in the USSR in the 1920s. The result was a control programme involving the removal of 40,000 to 50,000 wolves a year. As late as 1979, 32,000 were killed in the USSR. Despite this loss, wolves survive as a viable population in the USSR, where they are estimated to number about 70,000. Elsewhere in Eurasia farmers have largely wiped out local wolf populations, with increasing success as the human population spread and forests were converted to agriculture.

In contrast to North America, where there are no records of wolves killing people, deaths from wolf attacks are well-documented in parts of Eurasia. In some cases, the wolves involved were rabid, while in others it is suspected that they had been cross-bred with domestic dogs. In general, wolves are not dangerous to humans.

Grey wolves are still found in small numbers in parts of India. Known as pale-footed wolves, they are considerably smaller than the timber wolf of northern regions, although they are the same species. Diminishing habitat, human pressure and persecution are responsible for their low numbers. There are occasional reports of these wolves taking children. On the other hand, they have been reported to bring up abandoned children, or 'wolf boys', a tradition that inspired Rudyard Kipling's story of Mowgli in the *Jungle Book*.

The **red wolf**, found only in the USA, is smaller than the grey wolf. It usually has a cinnamon and tawny coat with grey and black highlights, but can be black. Its habitat consists of swamps and wetlands, and it used to be found throughout the southeastern USA from southern Florida to central Texas. It became extinct in the wild by the early 1980s. However, a captive breeding programme has made it possible to reintroduce the red wolf in South Carolina. The red wolf feeds on swamp rabbits and nutria (a species introduced into the United States) (also called coypu). It lives in packs like the grey wolf.

Despite its name, the **maned wolf** is not closely related to the true wolves. It equals them in size, but it is more lightly built. It has extremely long legs, long bushy, reddish-yellow fur, and an erect mane. The muzzle

The red wolf (Canis rufus) *is smaller, and much rarer, than its grey cousin.*

The maned wolf (Chrysocyon brachyurus).

and legs are dark, sometimes almost black. The tail is usually dark, but may be light, even white. Monogamous pairs share a territory, but they do not hunt together. Maned wolves live in grassland and scrubland, as well as agricultural areas in central and eastern South America. Almost half their food consists of vegetable matter, the rest being small mammals, birds, fruit and insects. They are known to take chickens and, occasionally, newborn lambs and piglets. Very rarely, the maned wolf may eat carrion. Two pups are usually born after a gestation period of 62-66 days, but the litter may number up to five individuals. The nest is in thick brush, and the den a crevice between rocks. The maned wolf's fur and flesh have no value. It is widely killed as a pest.

The **coyote**, sometimes called the prairie wolf, is found from southern Canada through most of North America, and into northern Central America. It has a grizzled, buff-grey coat, with brownish or yellow outer ears, forelegs and feet. The throat and belly are light grey to white. A grey-black band sometimes shows on the mid-back. After mating in January-March, a litter of between 6 and 18 cubs is born in a den, which may have originally belonged to badgers, skunks or woodchucks. Apart from a wide range of small mammals, birds and insects, coyotes even kill animals as large as pronghorn antelopes, deer and mountain sheep. Predation on livestock and poultry has led to persecution, but coyotes have, nevertheless, more than held their own and have increased in numbers and range. ■

The coyote (Canis latrans) *seen here in Jasper National Park, Canada has managed to maintain its numbers despite extensive persecution by man.*

Foxes 4

The red fox is now the most widespread wild dog, found from the Arctic Circle to North Africa, central America, and the Asiatic steppes. It was introduced by European settlers into eastern USA in the 17th century and also to Australia. There is a wide variation in size, but European specimens are about the size of a domestic cocker spaniel, with a large, white-tipped bushy tail. The red fox may range from greyish to rust-red to flame-red, but it is usually reddish-brown. The backs of the ears are black, as are often the lower limbs.

The red fox (Vulpes vulpes) *is not only the most widely distributed member of the Canid family, it is probably the most adaptable and successful of all carnivores.*

Red foxes live in monogamous pairs instead of in packs like wolves. The pair usually mates in late winter or early spring, and the young are born after 49-55 days. Both parents care for the pups, which remain in the den for the first 3 months. By the end of the first year, the young will have left their parents and dispersed.

Red foxes are adaptable to many different habitats, from Arctic tundra to the very centre of cities, such as London, Paris and Stockholm. They live in mixed landscapes, including woodlands, uplands, deserts, sand dunes and farmland. Their prey is extremely varied. Rodents and rabbits, birds, beetles and earthworms all serve as food. Red foxes also eat fruit, as well as garbage and carrion. Surplus food may be cached, and foxes remember well the location of their hoards.

In spite of heavy losses due to trapping, and predator and rabies control, red fox populations are highly resilient. Populations decimated by rabies usually recover quickly.

Little is known of the **corsac fox**, which ranges from Afghanistan through south-central Asia and Siberia, to northeast China. It closely resembles the red fox, but is more sturdily built. It is grey to grey-red in colour with a white chin. It lives in steppes and semi-desert, and avoids agricultural areas. Several individuals may share dens in burrows. Small packs, which may have been family groups, have been seen hunting.

In North America, the **swift** or **kit fox**, is the native relative of the red fox, which was introduced. It originally inhabited the western half of the continent from southern Alberta and Saskatchewan in Canada to northwestern Texas. It is being reintroduced into Canada, where it became extinct, and is no longer found in many parts of the USA. The swift fox is much smaller than the red fox, and has a greyish back and buff-orange underbelly in winter, with the upperside rather redder in summer. Its black-tipped tail is very bushy. It is active in the evening, during the night and early in the morning, when it feeds on small mammals, especially rabbits, birds, amphibians and insects. The swift fox mates in December-January and gives birth to 3-6 pups. Although swift foxes have lived for up to 13 years in captivity, their life expectancy in the wild is thought to be as little as 3-4 years.

The **Bengal fox** of the Indian sub-continent is smaller than the red fox and has a sandy-orange coat with a black-tipped tail. It is believed to live in monogamous pairs, but it is a solitary hunter of small mammals, ground-nesting birds and insects. Its habitat is open country, thorny scrub and semi-desert in the Indian plains. It is reported to be trapped in southern India for the supposed medicinal properties of its flesh.

The **Tibetan fox** lives in high mountain steppes on the Tibetan plateau and its border areas in India and Nepal. It is nearly as big as the red fox, but lighter. The coat is tan-to-rusty in colour, with grey sides and rump. The tip of the tail is white. Little is known of the Tibetan fox's life, but it feeds on rodents, rabbits, hares, and ground birds, and shows a particular liking for the black-lipped pika (a mouse hare). Tibetans trap these foxes for their fur, which is used to make hats.

Cape foxes (Vulpes chama) *are nocturnal solitary hunters feeding mainly on rodents, insects, and fruit. They are confined to southern Africa.*

A small fox in southern Africa is the **Cape fox**, also called the 'silver jackal'. The Cape fox is grey or silver-grey with a long bushy tail, half the length of the body. The head, ears and upper forelimbs are reddish. There are white patches on the cheeks and dark-brown or black patches on the thighs, while the throat is buffy and the underparts nearly white. Although they are rarely seen in groups, Cape foxes have been found to have communal dens in the southern Kalahari. They are mainly nocturnal, solitary hunters, feeding on rodents, rabbits, insects, beetle larvae and reptiles, as well as fruit. They will take lambs, and are therefore killed by farmers.

The **grey fox** belongs to a different genus (**Urocyon**) from the red and swift species (**Vulpes**). Its coat is mottled or grizzled, with a dark stripe down the back. It has a small black mane. Parts of the neck, flanks and legs are rufous, while the face and underbelly are white or buff. Once known from Canada to Venezuela in deciduous forests, the grey fox disappeared from Canada from about the time the red fox was introduced in the 17th century,

Portrait of a young male ***grey fox*** (Urocyon cinereoargenteus).

The Arctic fox (Alopex lagopus) *has a remarkable tolerance to cold conditions.*

but has since recolonised some parts of Ontario, Manitoba and Quebec. Prized for its fur, it is widely trapped. The usual litter size is about four, but may be as many as ten.

A smaller and rarer relative of the grey fox is the **island grey fox**, which is found only on the six largest Channel Islands off the coast of southern California.

The **Arctic fox** is little more than half the size of the average red fox. It has smaller, more rounded ears and a slightly shorter and broader muzzle. Most Arctic foxes are white, becoming grey-to-brownish-white on the back in summer. Some may have a blue sheen on a lightish-brown winter coat, which becomes chocolate-brown in summer. Although they are distributed throughout the range, blue Arctic foxes become more common to the west in Alaska and on islands. Mating occurs in spring and cubs are born 51-54 days later. Litter sizes range from 6-16, but as many as 19 have been recorded on Wrangel Island, off northeast Siberia, when lemmings were plentiful as prey. Some dens, which have complex burrow systems, are used repeatedly, with one estimated to have been occupied continuously for 300 years. Arctic foxes feed on ptarmigan, migrant wading birds, lemmings, seabird chicks and eggs, seal carcasses and some vegetable matter.

One of the smallest foxes is the **bat-eared fox** of Africa, which has large upstanding ears. The coat is grey-buff above, paler and more beige below. It has a black face mask, and black on the muzzle, ear tips, front legs, lower back legs and the last third of the tail. A black stripe runs along the back. Separate populations exist in southern Africa, and in east and northeast Africa in open grasslands, semi-desert and bush country. Bat-eared foxes live in pairs, accompanied by the young of the year. Their diet consists almost exclusively of termites and beetles. Lizards and mice may be taken, but this fox has largely abandoned mammal prey. ■

The appropriately named ***bat-eared fox*** (Otocyon megalotis) *is mainly insectivorous.*

Desert Foxes 5

North Africa and the Middle East, from Lebanon and Israel to Pakistan, are the home of the desert foxes. All have small bodies, thick fur and large ears which enable them to pick up the sounds of prey. They are active at night, like most desert animals, and very little is known about them.

The Fennec fox (Fennecus zerda) *has particularly large ears enabling it to locate its prey at night, and heavily furred paws which help to insulate its feet from the hot desert sand.*

The **Fennec fox** is the smallest of all the wild dogs, with a maximum body length of 41cm, plus tail of up to 31cm. Its prominent ears are 15cm in height. It weighs 1-1.5kg. The soles of the feet are furred, providing them with insulation and grip in the hot sandy deserts. Groups of up to 10 have been seen, but their relationship has not been established. Females defend the nest site, and the males do not enter the den. Mainly nocturnal, the fennec is omnivorous and eats plants as well as rodents, birds, eggs, reptiles and insects. Although it has been sighted in Sinai, its established range is from Egypt to Morocco and the Sahara.

In the Sahel region, along the southern edge of the Sahara as far east as the Red Sea, the **pale fox** lives on berries and fruit, sometimes taking birds, small rodents and reptiles. Similar in body structure to the red fox, the pale fox is smaller and has a pale fawn body and ears, with a reddish-brown, black-tipped tail. Little is known about this fox, which has been seen in pairs and in groups presumed to be family parties.

Biologist examines a live-trapped ***Blanford's fox*** (Vulpes cana) *before releasing it.*

Blanford's fox is about the same size as the fennec, but it is blotchy black, grey and white, often with a dark tip to the bushy tail. The appearance is cat-like, as are its gait and demeanour. This fox is resident from Israel to Baluchistan, Afghanistan and southern Turkestan. The trapping of specimens in Oman suggests that it is also found throughout the Arabian peninsula. It prefers cliffs in mountainous regions as habitat. A solitary hunter, it feeds on ripe melons, seedless grapes and Russian chives, as well as insects, including locusts. Blanford's fox is easily trapped and has been heavily exploited for its fur in its countries of origin. It is not known in international trade.

Ruppell's fox, slightly larger than the pale fox, is a light sandy colour with black patches on the face. The tail has a conspicuous white tip. The fur is soft and dense. It has a wide range from Morocco to Afghanistan, including the Sahara and Arabian deserts. It is thought to live mainly on insects, and may be gregarious, groups of 3-15 having been seen sunning themselves in the evening before hunting. ■

Little is known about ***Ruppell's fox*** (Vulpes rueppelli) *which is another desert species.*

Zorros (foxes) of South America 6

Many of South America's animals and plants are found nowhere else in the world. This is true of nine of the ten wild dogs. The tenth is the grey fox, which extends from Venezuela through Central America to the southern USA. The continent has six species of wild dog, which are often called 'foxes', although they are not true foxes in the Vulpes genus, like the red fox. They belong to the genera Dusicyon or Cerdocyon. To distinguish them from the true foxes, the name 'zorro', the Spanish for fox, is preferable. Very little is known about the way of life of most of them.

The zorros are found in every habitat on the continent, from the coastal deserts (Sechuran zorro), to the open savanna (grey zorro, culpeo, Azara's zorro). The small-eared zorro is unique among canids in living in the rain forest, while the crab-eating zorro inhabits coastal and lowland forests.

The **crab-eating zorro** is about the same size as the red fox. It is found in savannas and woodlands from Colombia and Venezuela to northern Argentina. The coat is grey-brown on the back; the face, ears and legs are rufous, and the throat and underbelly white. The ear tips, back of the legs and the tip of the tail are black. Crab-eating zorros form monogamous pairs, but they hunt alone for rodents, reptiles, birds, insects, fruit and, of course, crabs.

Crab-eating fox (Cerdocyon thous) *Venezuela, with typical long pointed skull and large ears.*

Culpeo (Dusicyon culpaeus), *Andes.*

The **culpeo** is also called the South American red fox, although the coat is grizzled-grey on the back and shoulders. The head, neck, ears and legs are tawny and the tip of the tail is black. It is larger than the red fox. The culpeo feeds on rodents, rabbits, hares and berries. It also has a reputation for predation on livestock, especially lambs. Thus it is killed as a pest, and it is also hunted locally for its fur. The mountains and pampas of the Andean and Patagonian regions are the culpeo's habitat. The social behaviour is unknown, but litters of 3-8 cubs have been recorded in Chile.

Bush dog (Speothos venaticus).

The **grey zorro** is found from southern Peru to Patagonia in plains, pampas, deserts and low mountains, where it feeds on rodents and berries. About the size of a red fox, the grey zorro has a brindled grey coat, paler grey underparts, and a rust-coloured head, flecked with white. There is a black spot on the chin. Little is known about its behaviour. Considerable trade exists in grey zorro skins, mainly from Argentina to western Europe.

Azara's zorro is found in the pampas, hills and deserts in Argentina, Uruguay, Paraguay and southeast Brazil. It is similar in colour to the grey zorro, but larger. It has large ears. Azara's zorro lives in monogamous pairs, and the cubs join them in hunting rodents, rabbits, hares and birds. It is heavily hunted for its fur.

The **small-eared zorro** is dark grey to black on the back, while the underside is rufous, mixed with black, white and grey. The tail has thick black hair. The forests of the Amazon basin are the home of the small-eared zorro.

The open grasslands in the sandy coastal semi-desert of northern Peru and southern Ecuador are the home of the **Sechuran zorro**. It is the least known wild dog. About the size of a red fox, this zorro's coat is a pure, pale agouti colour with a black-tipped tail.

The **hoary zorro** is small. Grey upper parts combine with a cream or fawn underside, the ears and outside of the legs being rufous. It has a short muzzle, small teeth, and a dark stripe on the upper side of the black-tipped tail. It lives in the upper mountain areas of open woodland and bush in south-central Brazil.

South America is also home to the **bush dog**, which is small and resembles a stocky weasel or civet. It has a broad face, with small ears, a short bushy tail and very short legs. The head and neck are reddish tan or tawny, becoming blackish towards the tail. The feet are webbed. Bush dogs are found in forest and wet savannas from Panama through South America as far south as northern Argentina. Little is known of the bush dog, except that it is the most social of the small canids, living in groups of up to ten. It is active by day, when it hunts rodents, including agouti, and even larger animals, such as capybaras, the largest rodents in the world, as well as rheas, relatives of the ostrich. ■

Jackals 7

Widespread from Burma through the Indian subcontinent, the golden jackal is also found in the Middle East and north and east Africa, and in southeast Europe. Bigger than the red fox, the golden jackal usually has a pale gold-brown or brown-tipped yellow coat. The back is black and grey and the head, ears and sides can be rufous. The underside is frequently ginger or nearly white, and the tip of the tail is black. Golden jackals usually live in mated pairs, but commonly form groups, which may be as large as 20. This jackal is omnivorous, eating small mammals, birds, reptiles, amphibians and carrion. It commonly scavenges the kills of larger carnivores, including tigers and lions. Its close relatives, resident only in Africa, are the side-striped and black-backed jackals, which are about the same size.

Black-backed jackal (Canis mesomelas) *chases vultures from kill, Kenya.*

The **black-backed jackal** is reddish-brown or ginger with a black saddle from the nape to the tail. The jackals live in pairs, and often hunt together. They have about 4 pups, which they move from den to den. The previous year's litter may help rear the new pups. They feed on small mammals, including small antelopes, birds, invertebrates, grubs and fruit. There are two separate populations, one in southern Africa, and the other in east Africa, where they prefer drier areas than the side-striped jackal. They are killed because they take sheep and for their fur, for local uses.

Simien jackal (Canis simensis), *Ethiopia.*

The **side-striped jackal** has a greyish-yellow coat with a white stripe from elbow to hip, which is not always visible. The tail has a white tip. These jackals live in pairs with their young, which disperse just before they are one year old. Their range is tropical African moist woodlands, although they are not found in rain forests. The Buganda tribe uses this jackal's heart as medicine to treat epilepsy.

In the Ethiopian highlands, the **Simien jackal** has also been called the Abyssinian wolf, or Simien fox. It is bright reddish in colour with white patches on throat, neck and chest and the basal half of the tail. The lower half of the tail is black. The muzzle is long and thin and the legs relatively long. Little is known about this animal. It is most frequently seen alone while foraging by day, but pairs also hunt together. Pairs and groups of up to nine congregate mornings and evenings with noisy greetings. Adults have been seen sleeping with sub-adults in a group. Parents are assisted in rearing the pups by helpers from previous litters. Food is regurgitated for the young. The Simien jackal's food consists of rodents, such as hares and giant mole rats. A jackal has been seen scavenging a cow carcass, and others have been observed hunting - unsuccessfully - reedbuck and mountain nyala calves. The species is found above 3,000 metres in the Ethiopian highlands, with the largest population (400-500) in the Bale Mountains National Park.

Wild dogs (Lycaon pictus) *squabbling over their kill, Southern Africa.*

The **African wild dog**, which has been extirpated in much of its former range in parts of north and northwest Africa, is the most distinctive of the whole canid family. Almost as big as the grey wolf, but lighter, it has a patchwork coat of yellow, black, white and grey blotches. The ears are large and rounded, the legs long, and the broad bushy tails usually have a white tip. African wild dogs live in packs composed of several related adult males, and from one to several related adult females originating from a different pack, and their pups. Mating occurs only between the dominant female and male and there is no fixed reproductive period. Usually, 7-10 pups are born after 69-73 days, but there may be as many as 19. All the members of the pack feed the mother and pups at the den by regurgitating food. When they are about 18 months old, the female pups disperse and join unrelated packs. Wild dog packs are nomadic, and their home range may show considerable overlap.

African wild dogs run down their prey, which includes large ungulates, such as wildebeest and kudu, as well as smaller reedbuck, gazelles and duikers. Like the dhole in Asia, the African wild dog is hated because of its apparent voracity and indiscriminate killing of game and livestock. Despite its role in natural ecosystems, it has been ruthlessly persecuted, and there have been calls for its extermination. As a result, it has disappeared from many parts of Africa and is rare even where it survives. It is one of the most endangered of all wild dogs.

Dholes (Cuon alpinus) *or Asiatic red dogs are now rare in many parts of their range.*

Packs of **dholes** or Asiatic red dogs are found in jungles in the Indian subcontinent and southeast Asia, forests and alpine areas in China and the USSR and as far east as Korea, as well as on the islands of Java and Sumatra. Larger and more 'dog-like' than the red fox, dholes have rounded ears and a long, moderately bushy, tail. The fur is evenly tawny or dark red with a slightly darker tail and lighter underparts. Dholes are highly social animals, which live in packs of 5-12. Three or more adults have been seen feeding pups and a lactating mother. The pack regurgitates meat for the young when they leave the den, as well as allowing them access to kills. Like wolves, dholes run down their prey, which includes large deer, biting as they run and feeding while the prey is still alive. As with African wild dogs, this behaviour has led to unjustified revulsion and to calls for their extermination. Persecution, along

Raccoon dog (Nyctereutes procyonoides), *with characteristic black facial mask, eating fish, Japan.*

with disease and poisoning by baits put out for wolves in the USSR, has resulted in their becoming rare in many parts of their range.

A curious-looking dog originating in eastern Asia, including Japan, is the **raccoon dog**, which, as its name implies, resembles a raccoon. Introduced to the western USSR for fur farming, it escaped into the wild and has spread westwards into Germany, Czechoslovakia and Romania. Raccoon dogs have long, brindled black-brown body fur with a black facial mask, sleek black legs and black fur on the upper side of the tail. In winter they grow thick pelts with hairs up to 120mm long. Little is known about the raccoon dog's social life, but it is believed to live in pairs or family groups, and to forage mainly at night in thick undergrowth. It is the only wild dog which hibernates. The litter consists of 4-6 pups, but there may be as many as 12. Although in zoos or ranches they may live for up to 11 years, in the wild the lifespan may be as little as 3-4 years. ■

Dingos 8

Australia has no true indigenous wild dogs. The red fox was introduced by European settlers. The well-known dingo was considered by early settlers to be a wild dog, but it is now generally accepted that it descended from domesticated dogs which were brought to Australia by the Aborigines from the southeast Asian islands 3,500 to 4,000 years ago. It became wild or 'feral'. Feral dogs of southern Asia are now considered to be more closely related to the dingo than to the domestic dog. It has been argued too that the dingo and its close relatives are descended from the Indian pale-footed wolf.

Dingos in Australia stand 40-65cm at the shoulder and weigh up to 21kg. Their relatives in southern Asia are smaller. The coat is usually ginger, but varies from a light sandy colour to a deep red-ginger. There are irregular white markings on feet, chest and the tip of the tail, which is long and bushy. Dingos can be distinguished from the domestic dog by their larger canine and carnassial teeth, differences in skull bones, and the pattern of breeding. Domestic dogs breed throughout the year, but dingos breed only once, in March-April (autumn to early winter in Australia) and August-September in Thailand and probably the rest of southern Asia.

The dingo (Canis familiaris dingo), *Northern Territory, Australia. Dingos probably descended from domesticated dogs brought to Australia by the early Aborigine settlers.*

Because of their predation on sheep, ***dingos*** *have been subject to extensive control by man and have been virtually wiped out in the sheep-farming areas of east and west Australia.*

Dingos in Australia may be found in deserts and deep forest, but they also favour forest edges, heath and grassland. They usually live in packs, which hunt and breed together like wolves, dholes and African wild dogs. About five pups are born after 62 days' gestation, but, within the pack, the dominant female kills the young of other females.

Dingos *usually live together in packs which hunt and breed together like wolves.*

Kangaroos and wallabies make up most of the diet of dingos, but lizards, rabbits, rodents and geese are also eaten, as are fruits and arid-land plants.

Because of their predation on sheep, dingos have been heavily killed in predator control programmes; bounties are still paid for skins and scalps. They have been virtually wiped out in the sheep - grazing areas of east and west Australia. But, despite this decimation, the dingo population has thrived elsewhere. However, it is threatened by interbreeding with domestic dogs.

In Asia, dingo-like feral dogs live in close association with people. They eat food scavenged from humans, including rice. In parts of southeast Asia, these dogs are killed for food. Although little research has been done on the effect of interbreeding with domestic dogs in Asia, it is considered a principal threat to the survival of this form. ■

Wild Dogs and People 9

Dogs and cats have been favourite pets and companions of people for thousands of years. There is little doubt that dogs were the first to be domesticated, almost certainly from wolves. Wolf packs probably took advantage of the kills of Stone Age hunters, who, in turn, stole meat from the wolves. Young wolves may have become pets and assisted in locating and killing game. The earliest evidence of clearly domestic dogs dates from 11,000 years ago in Iran and 9,500 years ago in Yorkshire, England. They were ancestors of the vast number of domestic dogs of over 400 breeds found throughout the world today, dwarfing the populations of both their wolf ancestor and all other wild dogs.

Sculptures and murals of the Ancient Egyptians depict the African wild dog, now extinct north of the Sahara, as well as breeds of domestic dogs.

While wild dogs have not figured in heraldry to the extent of the lion and tiger, they have been more prominent in folklore. The founding of Rome by Romulus and Remus, who were reputedly suckled by a she-wolf, is one of the most famous legends. Reports of children raised by wolves have occurred through the ages, even up to the present time in India. The impact of Rudyard Kipling's tale of Mowgli living with the wolves of Seoni, in central India, was so great that it was used to organise boys as 'wolf cub packs' before becoming scouts.

Some 2,500 years ago, Aesop, a slave of the Greeks, incorporated the wolf in his fables. Particularly well-known is the story of the shepherd boy who cried "Wolf!" to enjoy the spectacle of the excitement it aroused. After false alarms, everyone assumed he was playing a game, and so they did not react when he really faced a wolf.

Were-wolves - people able to change into wolves and carry out attacks - are common legends in many societies.

The power and hunting ability of the wolf have inspired men to adopt its name. During the Second World War, fleets of German U-boats were described as 'wolf packs'.

The Brothers Grimm's fairy tale of Little Red Riding Hood and the wolf which ate and took the place of her grandmother, and then threatened her, has probably enthralled and frightened more children than almost any other animal story.

Competing with the wolf in folklore is the red fox, usually renowned for its intelligence and cunning. Again, it was Aesop who made the fox a leading character in his fables. His tale of the fox which described some juicy grapes it could not reach as 'sour', and said he did not want them anyway, gave rise to the common expression 'sour grapes' to describe similar situations in daily life.

'Reynard' means 'fox' to most people. Its origin is over 1,000 years old, and the stories of Reynard the Fox appear in many European languages. Reynard, who had equally well-known companions in Bruin the Bear and Chanticleer the Cock, is wily and cunning.

Foxes also feature in Japanese folklore. There is even a temple dedicated to the Fox Goddess, Inari no-jinja, in Kyoto.

Today human settlement has moved deep into the former habitat of the wild dogs. In turn, some species have now moved into the settlements. Foxes have become quite common urban animals, even living in central London and other great cities. Coyotes make their home in American cities and towns, and jackals still inhabit Delhi.

For centuries, wild dogs have been treated both as pests and an exploitable resource. They are blamed and killed for destroying poultry, livestock and valuable wildlife, and causing disease in people and animals. As a resource, they are killed and their flesh and fur used for a variety of purposes.

Trade in wild dog furs

The international trade in fur involves animals which are both farmed, and taken from the wild. Countries on every continent are both producers and consumers of canid fur products. The great majority of the furs used in international commerce comes from farmed animals - the red fox, the Arctic fox, and the raccoon dog are intensively farmed in many countries.

***Silver fox** coat.*

Of the 34 wild species, 20 are used by man for their fur, mostly for local use. Most of those killed for their fur are in no danger of extinction, although local populations could disappear if adequate controls are not enforced by the responsible governments. The red fox, grey fox, coyote and Arctic fox, are all widespread and abundant. In all of North America, nearly 1.5 million wild canids were killed for their pelts in the 1982/83 trapping season. But even this rate of harvest does not appear to threaten wild populations. Harvests of grey wolves in Canada, the Soviet Union, and Alaska pose little threat. However, the effects of a legal trade on other less stable populations need to be assessed.

All parts of the wolf's body are traded, including feet, teeth, claws and flesh. The great majority of wolf exports, however, is unfinished skins, sent mainly to the United States and Great Britain.

There is a large international trade in the grey zorro. Most of the furs go to the Federal Republic of Germany from Argentina and, to a lesser extent, from Chile. Culpeo skins also go to Germany from Argentina. International trade in other wild harvested species is very limited.

Controlling predation

Killing of wild dogs to control predation on livestock is widespread. While it is possible to show that the overall impact of livestock predation is small, it can be locally heavy due to extreme reductions of normal prey. In such cases the loss of livestock can be critical to the livelihood of small farmers. At the same time, it must be realised that wild dogs are often blamed for killings by domestic dogs. The National Farmers Union in Britain stated recently that 10,000 sheep had been killed by rampaging domestic dogs in one year. The coyote appears to be sufficiently prolific to survive in good numbers despite widespread poisoning, but action against the rare African wild dog is a major contributor to its drift towards extinction. Wolves are prime targets of hunters and farmers, in many cases with little or no justification. In Norway, the killing of a few rare individuals provoked jubilation in local communities.

Control operations involving the red fox have been undertaken in recent years in Europe because of the spread of rabies. New techniques of inoculation have been developed. Treated chicken heads and other baits scattered in the fox's habitat have made it possible to reduce poisoning, which can threaten many other species, and even children. However, poisoned baits are frequently used elsewhere, and when put out for wolves are said to be responsible for the decline of the dhole in the USSR. ■

Early methods of controlling fox populations frequently caused considerable suffering.

Priorities in Conservation of Wild Dogs 10

The Canid Specialist Group has recommended the following priorities for conservation of threatened species:

Red wolf - After extermination in the wild in the 1970s, the red wolf has been bred in captivity and reintroduced in North Carolina. A second site for reintroduction should be found and restocked as soon as possible.

Simien jackal - This is the most endangered canid in Sub-Saharan Africa and is limited to the Ethiopian highlands, where it numbers fewer than 1,000, mostly in the Bale Mountains National Park. Larger populations should be established outside the park.

African wild dog - Probably the most endangered large carnivore in Africa, already extinct in 19 of 32 countries in which it was formerly known. Only in Botswana, Ethiopia, Kenya, Tanzania, Zambia and Zimbabwe are there viable populations, which could survive for more than a decade. Complete protection is essential and should be combined with an education programme for livestock owners and children. Research into methods of minimising conflict with livestock owners is essential.

Grey wolf - The species is threatened with extinction in much of its range through loss of habitat, hunting, inter-breeding with domestic dogs, and human persecution. Protection, status surveys, education, and reintroduction are needed. Priority areas are Mexico, southwest and southeast USA, Michigan and Wisconsin, central Europe, northern Rockies, Norway, Sweden, Middle East and southwest Asia, and India.

Bush dog - The bush dog is believed to be rare in its range in South America. Little is known about it. Status surveys are needed, as well as studies of its behaviour and ecology in the wild. Captive breeding programmes should be considered.

Dhole - The dhole is unlikely to survive outside protected areas. Unexplained declines have occurred in Nepal and the USSR. Burma may hold the most viable population. Status surveys are required, especially in Burma and the USSR. The use of poisons against wolves in the USSR, where dholes co-exist with them, should be banned.

Grey zorro - Harvests are uncontrolled and there is extensive trade in the grey zorro, although there is no reliable information on population density or distribution. Status surveys are required, especially in Chile and Argentina, and harvest quotas should be imposed until reliable data are obtained. Existing laws in Chile should be enforced.

Maned wolf - Agricultural expansion and conflict with people cause concern for the maned wolf, although its status appears to have improved slightly. Local governments in its range in South America should support farming practices which are compatible with wildlife use; education programmes should be coordinated and expanded, and reintroduction programmes encouraged.

Island grey fox - This fox is found only on the six largest Channel Islands off the coast of southern California. The US Navy's proposed elimination of feral cats on San Clemente and San Nicholas islands should be supported, and quarantine should be imposed where domestic dogs are present in order to prevent transmission of disease. ■

Conclusion 11

In terms of ethics, it is argued that the human species, which itself originated through natural evolution, has no right to eliminate other species. Both have a right to exist in this world.

Alongside this argument in favour of conservation, there is also an emotional view. In the words of Dr David MacDonald, Chairman of the SSC Canid Specialist Group:

"Perhaps it is insight into their individualism won through our association with domestic dogs; perhaps it is the ability of some of them to withstand the worst that man can throw at them; perhaps it is that we see some reflection of ourselves in their opportunism; or perhaps it is simply the transfixing elegance of their jaunty grace. Whatever the pot pourri of reasons that blend into their appeal, the 34 species that comprise the family Canidae are special. They are special because they have, whether perceived as friend or foe, preoccupied the imagination of people for millennia. They are special because the breadth of their adaptations makes them enthralling to science. They are special because the contradictory facets of their relations with people perplex the conservationist. The possibility that we are heedlessly, perhaps needlessly, mismanaging many of them, is saddening; the probability that our negligence will force several more to extinction, should fill us with bottomless dismay".

There is no simple formula for conserving the wild dogs. Variations in their behaviour and circumstances make it impossible. The Simien jackal appears to need seclusion from people and especially from domestic dogs, if it is to survive in its highland refuges. The African wild dog's intolerance of human intrusion and its need for vast space mean certain conflict with modern man. The bush dog is so rarely seen that nothing is known about its status, and it could be threatened by killing and loss of habitat. On the other hand, red foxes are successful in urban settings, and other canids, ranging from coyotes to crab-eating zorros, seem able to thrive amidst human settlements.

Conservation of natural habitat must have the highest priority. This applies not only to wild dogs, but to all wild animals and plants. Destruction of natural habitat can only lead to diminishing wildlife populations and extinction of many species. However, nature conservation cannot be treated as an isolated programme. It must be accompanied by provision for the world's human population, which will continue to expand for at least several decades. Economic development will inevitably lead to conversion of natural areas to agriculture and other income generating purposes. It is the duty of planners and executives to see that damage to these areas is limited and that provision is made for the wild species to continue to exist there. Wildlife reserves of adequate size to support viable wild populations are of major importance. Where their size has to be limited, the needs of species can be provided for by creating corridors between reserves. Buffer zones, where wildlife co-exists with people, are desirable. Human activities in these zones should be compatible with wildlife and reinforce conservation objectives. There may be some conflict, but it is the task of wildlife managers to keep it to acceptable levels. Those who suffer loss and damage from abiding by the rules should receive reasonable compensation.

For the past 30 years, IUCN - The World Conservation Union has categorised threatened species. Those in danger of extinction are classed as 'Endangered'. Those likely to be endangered if present trends continue are

'Vulnerable'. In addition, there are categories for rare species, and for those about which so little is known that their status is 'Indeterminate' or 'Insufficiently Known'. Of the canids, the Simien jackal and the red wolf are classified as 'Endangered'; African wild dog, bush dog, dhole, grey wolf and maned wolf as 'Vulnerable'; and the Bengal fox is 'Insufficiently Known'.

In the light of recent surveys, the Canid Specialist Group has recommended that the African wild dog be listed as 'Endangered', and the grey zorro as 'Vulnerable'. The species which should be classed as 'Insufficiently Known' are Blanford's fox, corsac fox, fennec, hoary zorro, pale fox, Ruppell's fox, Sechuran zorro and small-eared zorro.

Poorly-conceived management programmes for various canids (largely prompted by rabies control, and attempts to limit predation on domestic stock) lead to the killing of hundreds of thousands of foxes annually, thousands of wolves, and a handful of African wild dogs. Lack or poor enforcement of regulations controlling hunting for commercial purposes (eg fur and food) also threaten some local populations. While the foxes and wolves appear to be able to withstand this onslaught, there are few African wild dogs left.

Government controls to prevent excessive hunting and trapping need to be strengthened and enforced so that wild populations are not driven to extinction. The trade in canid fur is both national and international; countries on every continent are both producers and consumers of Canid fur products. Most of the species wild-harvested for their fur do not appear to be in danger of extinction. The red fox, grey fox, coyote and Arctic fox are all widespread and abundant. Most canid furs used in commerce come from farmed animals because of the superior quality of the pelt. In those countries where there are wild harvests, it is essential that population sizes be ascertained, sustainable harvest levels be established, and annual monitoring of the population be implemented.

International commerce in wildlife is covered by the Convention on International Trade in Endangered Species of Wild Fauna and Flora (CITES). Species which are prohibited in trade are listed in Appendix I of this treaty which has been adopted by 109 countries. Currently the red wolf and the bush dog, as well as the grey wolf (with the exception of the populations in Canada, Alaska and the USSR) are listed on Appendix I. Appendix II includes species which are allowed in international trade; however such trade requires permits from the government in order to monitor the offtake because the species listed might become endangered if trade is not controlled. The dhole, maned wolf, Blanford's fox, grey zorro, Azara's zorro, culpeo and fennec are all listed on Appendix II.

Rabies is closely associated with foxes, although other carnivores may also carry and spread the disease. The challenge of the wave of rabies which has swept across Europe in the past 40 years has led to important advances in control, especially oral vaccination of foxes by spreading treated baits.

For many endangered species, captive breeding and subsequent reintroduction may offer a solution. Apart from the red wolf, captive-bred kit foxes are now being reintroduced to Canada. To be successful, a substantial number of founding animals with a broad genetic base needs to be established before numbers reach dangerously low levels in the wild.

For centuries, wild dogs have been treated both as pests and as an exploitable resource. They are blamed and killed for destroying poultry, livestock and valuable wildlife, and causing disease in people and animals. As a resource, they are killed and their flesh and fur used for a variety of purposes. ■

Distribution of Wild Dogs and their Relatives 12

Common name	Scientific name	Geographical distribution
Coyote	*Canis latrans*	Holarctic
Culpeo	*Dusicyon culpaeus*	South America
Dhole	*Cuon alpinus*	South & Southeast Asia & Australia
Dingo	*Canis familiaris dingo*	South & Southeast Asia & Australia
Dog, bush	*Speothos venaticus*	South America
Dog, raccoon	*Nyctereutes procyonoides*	Holarctic
Fox, Arctic	*Alopex lagopus*	Holarctic
Fox, Bengal	*Vulpes bengalensis*	South & Southeast Asia & Australia
Fox, Blanford's	*Vulpes cana*	North America & the Middle East
Fox, bat-eared	*Otocyon megalotis*	Sub-Saharan Africa
Fox, Cape	*Vulpes chama*	Sub-Saharan Africa
Fox, corsac	*Vulpes corsac*	South & Southeast Asia & Australia
Fox, Fennec	*Fennecus zerda*	North America & the Middle East
Fox, grey	*Urocyon cinereoargenteus*	South America, Holarctic
Fox, Island grey	*Urocyon littoralis*	Holarctic
Fox, pale	*Vulpes pallida*	North America & the Middle East
Fox, Ruppell's	*Vulpes rueppelli*	North America & the Middle East
Fox, red	*Vulpes vulpes*	Holarctic, North America & the Middle East, South & Southeast Asia & Australia
Fox, swift, or kit	*Vulpes velox*	Holarctic
Fox, Tibetan	*Vulpes ferrilata*	South & Southeast Asia & Australia
Jackal, black-backed	*Canis mesomelas*	Sub-Saharan Africa
Jackal, golden	*Canis aureus*	Sub-Saharan Africa, Middle East, Near East, Southern Europe, North America & the Middle East, South & Southeast Asia & Australia
Jackal, Simien	*Canis simensis*	Sub-Saharan Africa
Jackal, side-striped	*Canis adustus*	Sub-Saharan Africa
Wild dog, African	*Lycaon pictus*	Sub-Saharan Africa
Wolf, grey	*Canis lupus*	Holarctic
Wolf, maned	*Chrysocyon brachyurus*	South America
Wolf, red	*Canis rufus*	Holarctic
Zorro, Azara's	*Dusicyon gymnocercus*	South America
Zorro, crab-eating	*Cerdocyon thous*	South America
Zorro, grey	*Dusicyon griseus*	South America
Zorro, hoary	*Dusicyon vetulus*	South America
Zorro, Sechuran	*Dusicyon sechurae*	South America
Zorro, small-eared	*Dusicyon microtis*	South America

Acknowledgements:

Picture credits: *Key: t top b bottom c centre*
2 Jane Burton/Bruce Coleman, 3 E&P Bauer/Bruce Coleman, 4t Rick Weyerhauser/WWF, 4b Stephen Krasemann/NHPA, 5 Sue Boulton/ICCE, 6 G. Lacz/NHPA, 7 Mark Boulton/ICCE, 8 Hans Reinhard/Bruce Coleman, 9 E&P Bauer/Bruce Coleman, 10t Peter Jackson, 10b Eric Dragesco/WWF, 11 Mark Boulton/ICCE, 12 Jill Sneesby/Bruce Coleman, 13 Rod Williams/Bruce Coleman, 14t Andy Purcell/ICCE, 14b Mark Boulton/ICCE, 15 J Cancalosi/Bruce Coleman, 16t Steve Kaufman/Bruce Coleman, 16b I M Lindsay, 17t James Malcolm, 17b D&R Sullivan/Bruce Coleman, 18 David Macdonald, 19 Mark Boulton/ICCE, 20t J Stephenson/WWF, 20b Anthony Bannister/NHPA, 21 Peter Jackson, 22 Orion/NHPA, 23 Dr Laurie Corbet, 24t&b Dr Laurie Corbet, 25 Warwick Goble/Mary Evans Picture Library, 26 Jane Burton/Bruce Coleman, 27 International Fur Trade Federation, 28 Alain Compost/ICCE.

Text: Peter Jackson/Olga Sheean-Stone